Setting Priorities for Space Research

An Experiment in Methodology

TASK GROUP ON PRIORITIES IN SPACE RESEARCH

JOHN A. DUTTON, Pennsylvania State University, *Chair*
PHILIP ABELSON, American Association for the Advancement of Science
WILLIAM P. BISHOP, Desert Research Institute
LAWSON CROWE, University of Colorado
PETER DEWS, Harvard Medical School (retired)
ANGELO GUASTAFERRO, Lockheed Missiles and Space Company, Inc.
MOLLY K. MACAULEY, Resources for the Future
THOMAS A. POTEMRA, Johns Hopkins University
ARTHUR B.C. WALKER, JR., Stanford University

JOYCE M. PURCELL, Executive Secretary
CARMELA CHAMBERLAIN, Administrative Assistant

SPACE STUDIES BOARD

CLAUDE R. CANIZARES, Massachusetts Institute of Technology, *Chair*
LOUIS J. LANZEROTTI,* AT&T Bell Laboratories, *Former Chair*
JOHN A. ARMSTRONG, IBM Corporation (retired)
LAWRENCE BOGORAD, Harvard University
JOSEPH A. BURNS,* Cornell University
JOHN A. DUTTON,* Pennsylvania State University
ANTHONY W. ENGLAND, University of Michigan
JAMES P. FERRIS,* Rensselaer Polytechnic Institute
DANIEL J. FINK, D.J. Fink Associates, Inc.
HERBERT FRIEDMAN,* Naval Research Laboratory
MARTIN E. GLICKSMAN, Rensselaer Polytechnic Institute
RONALD GREELEY, Arizona State University
BILL GREEN, former member, U.S. House of Representatives
HAROLD J. GUY,* University of California, San Diego
NOEL W. HINNERS, Lockheed Martin Astronautics Company
ROBERT A. LAUDISE,* AT&T Bell Laboratories
RICHARD S. LINDZEN,* Massachusetts Institute of Technology
JANET G. LUHMANN, University of California, Berkeley
JOHN H. McELROY, University of Texas, Arlington
WILLIAM J. MERRELL, JR.,* Texas A&M University
ROBERTA BALSTAD MILLER, Consortium for International
    Earth Sciences Information Network
BERRIEN MOORE III, University of New Hampshire
NORMAN F. NESS,* University of Delaware
MARCIA NEUGEBAUER,* Jet Propulsion Laboratory
MARY JANE OSBORN, University of Connecticut Health Center
SIMON OSTRACH, Case Western Reserve University
JEREMIAH P. OSTRIKER,* Princeton University
CARLÉ M. PIETERS, Brown University
JUDITH PIPHER,* University of Rochester
MARCIA J. RIEKE, University of Arizona
ROLAND W. SCHMITT, Rensselaer Polytechnic Institute (retired)
JOHN A. SIMPSON, University of Chicago
WILLIAM A. SIRIGNANO,* University of California, Irvine
JOHN W. TOWNSEND,* NASA (retired)
FRED W. TUREK,* Northwestern University
ARTHUR B.C. WALKER, JR.,* Stanford University

MARC S. ALLEN, Director

---

*Former member.

COMMISSION ON PHYSICAL SCIENCES, MATHEMATICS, AND APPLICATIONS

ROBERT J. HERMANN, United Technologies Corporation, *Chair*
STEPHEN L. ADLER, Institute for Advanced Study
PETER M. BANKS, Environmental Research Institute of Michigan
SYLVIA T. CEYER, Massachusetts Institute of Technology
L. LOUIS HEGEDUS, W.R. Grace and Company
JOHN E. HOPCROFT, Cornell University
RHONDA J. HUGHES, Bryn Mawr College
SHIRLEY A. JACKSON, U.S. Nuclear Regulatory Commission
KENNETH I. KELLERMANN, National Radio Astronomy Observatory
KEN KENNEDY, Rice University
HANS MARK, University of Texas, Austin
THOMAS A. PRINCE, California Institute of Technology
JEROME SACKS, National Institute of Statistical Sciences
L.E. SCRIVEN, University of Minnesota
LEON T. SILVER, California Institute of Technology
CHARLES P. SLICHTER, University of Illinois at Urbana-Champaign
ALVIN W. TRIVELPIECE, Oak Ridge National Laboratory
SHMUEL WINOGRAD, IBM T.J. Watson Research Center
CHARLES A. ZRAKET, MITRE Corporation (retired)

NORMAN METZGER, Executive Director

Preface

Finding an effective and reasonably objective method for setting priorities across scientific disciplines is something of a holy grail of science policy, and one whose urgency continues to grow. Three years ago, the Space Studies Board Task Group on Priorities in Space Research, under the leadership of John A. Dutton, presented a thorough analysis of the pros and cons of priority setting (*Setting Priorities for Space Research: Opportunities and Imperatives*, National Academy Press, Washington, D.C., 1992) and recommended that an effort to develop such a method proceed. The next phase involved this task group, and eventually the full Board, in an ambitious attempt to construct a formal, semiquantitative methodology to set priorities among major space science projects using both scientific and societal criteria. They also conducted a trial application of the methodology to a set of hypothetical space science initiatives. This is a report on the methodology and this exercise.

Like a great many worthy scientific experiments, this exercise did not yield exactly what the framers had anticipated. In particular, the Board was not able to reach a consensus on the task group's methodology for setting priorities. This report contains an analysis of general issues in priority setting and presents a valuable record of the strengths and weaknesses of the task group's proposed methodology, one among many possible approaches. This report will inform continuing efforts in this area, including a recently undertaken, congressionally mandated Board study. The latter study is following a different approach toward different, but related, goals—agency organization and technology utilization are being analyzed together with the research priority-setting problem. The results of this study are reported in *Managing the Space Sciences* (National Academy Press, Washington, D.C., 1995).

Although it was hoped that the methodology developed by the task group and presented here would solve the priority-setting problem, the result that more work is needed should neither surprise nor discourage. The Task Group on Priorities in Space Research worked long and hard on a problem that many would have shunned. By doing so, its members have given us this important contribution to science policy and have earned the gratitude of the Space Studies Board and of the broader community of researchers.

Claude R. Canizares, *Chair*
Space Studies Board

Contents

	EXECUTIVE SUMMARY	1
1	INTRODUCTION	3

    The Existing Framework for Priority Setting in
       Space Research, 5
    An Axiom for Scientific Research in Space, 6
    Rationale for Priorities in Space Research, 6
    Context for Priorities in Federally Supported Research, 7
    Taking Up the Challenge: A Method and an Experiment, 10
    Notes, 11

2   THE PROCESS OF SETTING PRIORITIES   13

    Characteristics of an Effective Priority-Setting Process, 13
    Key Elements of a Priority-Setting Process, 14
    A Schematic Sequence for Setting Priorities, 18
    Notes, 20

3   THE TOOLS: THE ADVOCACY STATEMENT AND
    EVALUATION FORM   22

    The Advocacy Statement, 22
    The Evaluation Form, 25

4	TESTING THE TOOLS AND METHODOLOGY	26
	Testing the Proposed Priority-Setting Process, 26	
	Detailed Results of the Trials, 27	
	General Conclusions from the Trials, 32	
5	BOARD ASSESSMENT OF THE TRIALS AND RESPONSE TO TASK GROUP RECOMMENDATIONS	34
	Board Meeting of February 1993, 35	
	Board Meeting of June 1993, 36	
	Board Meeting of November 1993, 37	
6	CONCLUSIONS	39

APPENDIXES

A	Advocacy Statement	45
B	Qualitative Evaluation Form	61
C	Quantitative Evaluation Form	65
D	Notes on Quantitative Evaluation Schemes	75

Executive Summary

In 1989, the Space Studies Board created the Task Group on Priorities in Space Research to determine whether scientists should take a role in recommending priorities for long-term space research initiatives and, if so, to analyze the priority-setting problem in this context and develop a method by which such priorities could be established. After answering the first question in the affirmative in a previous report, *Setting Priorities for Space Research: Opportunities and Imperatives* (National Academy Press, Washington, D.C., 1992), the task group set out to accomplish the second task.

The basic assumption in developing a priority-setting process is that a reasoned and structured approach for ordering competing initiatives will yield better results than other ways of proceeding. The task group proceeded from the principle that the central criterion for evaluating a research initiative must be its scientific merit—the value of the initiative to the proposing discipline and to science generally. But because space research initiatives are supported by public funds, other key criteria include the expected contribution to national goals (including the enhancement of human understanding), the cost of the initiative, and the likelihood of success. To be effective, a priority-setting process must also reflect the values of the organizations using it, be sensitive to political and social contexts, be efficient, and provide a useful product. The key elements of such a process include determining the categories of candidate initiatives that will be considered, specifying explicitly the criteria that will be used to evaluate initiatives, providing a mechanism for advocacy of initiatives, and providing a mechanism for evaluating the initiatives. Evaluation schemes can range from informal, subjective approaches to formal, quantitative methods. This general outline of a priority-setting process was expanded by the task group into a specific schematic sequence of distinct steps for selecting and ranking initiatives. The task group developed a two-stage methodology for priority setting and constructed a procedure and format to support the methodology. The first of two instruments developed was a standard format for structuring proposals for space research initiatives. The second instrument was a formal, semiquantitative appraisal procedure for evaluating competing proposals.

To help guide the development of the priority-setting process and instruments, the task group conducted two trials of preliminary versions. In the first, a small informal group of practicing scientists was convened at a workshop to evaluate a set of strawman initiative proposals prepared with the help of Board discipline committees. The results of this trial were used to refine the instruments. In a second trial, the Board itself exercised revised versions of the instruments and assessed their utility. The Board concluded that the method was not fully suitable for adoption on an operational basis. Reasons given by individual Board members for not supporting the proposed process included philosophical differences with the scope or approach of the method, reservations about the instruments themselves, and concerns about the ability of the Board and its committees to successfully implement them. Notwithstanding the reluctance of some members to adopt the proposed methodology, the Board remained broadly in accord with the task group's earlier finding that researchers should participate actively in priority setting.

This report makes available complete templates for the methodology, including the advocacy statement and evaluation forms, as well as an 11-step schema for a priority-setting process. From the beginning of its work, the task group was mindful that the issue of priority setting increasingly pervades all of federally supported science and that its work would have implications extending beyond space research. Thus, although the present report makes no recommendations for action by NASA or other government agencies, it provides the results of the task group's work for the use of others who may study priority-setting procedures or take up the challenge of implementing them in the future.

1

Introduction

The first report[1] of the Task Group on Priorities in Space Research argued that planning for the future of space research must be guided by two fundamental questions: *What should we do? How should we do it?* The report concluded that the answers to these questions required (1) setting priorities among space research initiatives and (2) developing a process for effectively implementing and managing the resulting agenda. The report argued that through use of an appropriate procedure, the space science community could create an ordering of major scientific space research initiatives, possibly in a multidimensional priority structure. Such an agenda would help the space agencies, the Administration, and Congress allocate resources and determine the appropriate place for space research activities among other national endeavors.

From the outset of its deliberations, the task group recognized that the context of federal support for space research, and for science generally, is changing rapidly. The end of the Cold War removed a major motivation for federal support of basic scientific research. Worldwide economic competition and a faltering economy at home have increased the demand for applications-oriented research and technological development. The twin pressures of the annual federal budget deficit and service on a steadily increasing national debt are expected to limit future federal research expenditures. As a result, the total cost of seizing all scientifically meritorious opportunities in space research exceeds the funds likely to be available in the foreseeable future. It is also clear that perceived relationships between major scientific initiatives and national goals are becoming increasingly significant in the allocation of federal research support.

For these and other reasons, the need to set priorities is widely recognized. Recent examples of processes or recommendations for priority setting include the following:

- Priorities for programmatic activities were developed by the former NASA Space Science and Applications Advisory Committee,[2] using a methodology created by its predecessor, the Space and Earth Science Advisory Committee.[3,4]
- A two-dimensional priority structure for scientific activities in the U.S. Global Change Research Program was developed and implemented by the Committee on Earth and Environmental Sciences of the Federal Coordinating Council for Science, Engineering, and Technology.[5]
- Recommendations for decadal priorities in research and infrastructural investments to advance astronomy and astrophysics were prepared by a special committee of the National Research Council.[6]
- A quantitative scheme for establishing priorities to select health technologies for detailed evaluation and assessment was proposed by a committee of the Institute of Medicine.[7]
- A variety of issues involved when local governments set priorities for capital improvement proposals were examined and various methodologies presented in a report of the Urban Institute.[8]

This report discusses some important considerations in attempting to develop and implement a priority-setting process for space research initiatives. As in the task group's first report, the goal is to address priorities among specific initiatives rather than to rank disciplines themselves or to answer the question of whether a balance should be maintained among disciplines.[9]

The task group also limited itself to considering research conducted in space. An excellent argument could be made for extending the priority-setting analysis to the issue of space-based versus ground-based science, especially in view of the large cost differential between space missions and typical ground-based research. From policy and legislative vantage points, however, space research is handled separately from conventional research, and an attempt to systematize comparisons across this boundary would need to explicitly address this reality, which was outside the task group's scope. The task group strove to restrict the scope of the effort to the immediate concerns of the Board, but was mindful that a successful methodology might constructively contribute to efforts to set priorities across broader ranges of the national scientific effort.

THE EXISTING FRAMEWORK FOR PRIORITY SETTING IN SPACE RESEARCH

The contemporary space research program has been constructed and managed in the context of broad statements of national commitment, advice from various sources on scientific strategies and general priorities, and a strategic plan on the programmatic level.

The National Space Act of 1958 established a commitment to space research that provides "expansion of human knowledge of the Earth and of the phenomena in the atmosphere and space." Presidential space policy directives have reinforced and amplified this commitment, stating that an objective of the U.S. civil space activities "shall be . . . to expand knowledge of the Earth, its environment, the solar system, and the universe . . . ."[10] Various national-level advisory groups have provided further recommendations, including the Advisory Committee on the Future of the U.S. Space Program, which argued that the civil space science program "should have first priority for NASA resources . . . ."[11] More recently, the Clinton Administration declared, ". . . the space program should create new knowledge that will contribute to our understanding of our environment and our place in the universe."[12]

Since the inception of the space program, the Space Studies Board has advised on the scientific merit and relative priorities of initiatives proposed by various disciplines in space science and applications. The Board has issued a series of disciplinary strategies assessing progress and recommending research directions and specific mission concepts that offer the most scientific benefit. In 1988, the Space Studies Board provided a comprehensive statement of research opportunities within space research disciplines for the period 1995 to 2015, but no attempt was made to establish priorities across the initiatives proposed by various disciplines.[13]

At the programmatic level, the former Office of Space Science and Applications (OSSA) issued a series of five-year strategic plans setting forth multidimensional priorities for both continuing efforts and requests for "new start" authority. These programmatic priorities were developed with the assistance of OSSA's internal scientific advisory committees.

There are two domains in which priority recommendations for space research are not now available from the scientific community. First, the recommendations of the Space Studies Board discipline committees are not combined into a comprehensive prioritized recommendation for all of space research; second, there is no comprehensive agenda for the period extending beyond the OSSA five-year plan.

AN AXIOM FOR SCIENTIFIC RESEARCH IN SPACE

The Space Studies Board charged the Task Group on Priorities in Space Research to determine whether or not a procedure for establishing priorities among space science initiatives was feasible and desirable, and if so, to develop a methodology to do so. In its previous report, the task group concluded that this would require that the objectives of scientific research in space be clarified and distinguished from other objectives of the civil space program. In order to set meaningful priorities over a wide range of activities, it is essential to have a clear vision of the real objectives of space research. While there are a variety of reasons for going into space, the task group's fundamental assumption was that the goal of scientific research in space is to improve our understanding of phenomena observed in, and from, space. To accomplish this goal, space research must acquire data in and from space and analyze and synthesize the information contained in that data. In brief, then, the acquisition of information, knowledge, and understanding are the objectives of scientific research in space.[14]

In this context, the task group concluded that the space science and applications communities should reach a consensus on priorities for scientific research initiatives in space. It argued that such priorities are desirable and necessary for the effective management of the nation's space research program and would enhance the contributions of these communities to meeting national goals.

RATIONALE FOR PRIORITIES IN SPACE RESEARCH

Lessons learned from the civil space program provide evidence to support this position. The past 35 years have been characterized by extraordinary accomplishments, opening the way for space research and human exploration of space for generations to come. Studies of the Earth from space, of the sun and planetary system, and of distant objects have revealed startling complexity and beauty in the biological and physical universe. In addition, much has been learned from past mistakes. This learning includes the relative merits of large and small missions, constraints on human activities in space, and operating principles that increase the probability of success. In particular, the space science community understands that the greatest scientific returns are achieved when scientific research objectives determine the nature of each mission. It also recognizes that congressional and public demands for accountability in our changing international and economic climate must be met.[15,16] The directive from the Senate appropriations subcommittee referred to in the preface to this report is evidence that the views of scientists about management of science and science priorities are valued by policymakers.

One result of progress in space research is a steadily growing list of proposed missions that promise to increase human knowledge and understanding. Proposals for large and small missions and for research projects in different fields—some scientifically mature, some still developing—are testimony to the vitality of space science.

In light of these considerations, the task group concluded that an orderly process is needed to determine which missions and other initiatives to pursue and in what order. As described above, it recognized from the beginning the imperative to consider national goals and national needs.

Because it is likely that present economic conditions will in part dictate which space research initiatives are ultimately pursued, a rational, fair, open, and orderly process for choosing among initiatives should take account of all important considerations. Examples of these considerations include achieving the greatest possible return on investment with respect to knowledge gained, obtaining knowledge relevant to technological development or economic improvement, and taking advantage of windows of opportunity. The task group proposed the following major criteria for evaluating initiatives:

- *Scientific merit*—the value of the initiative to the proposing discipline and to science generally;
- *Contributions to national goals*—specifically how the initiative will serve the nation;
- *Cost*—both project-specific and supporting costs, and also opportunity costs; and
- *Likelihood of success*—including timeliness, existence of a group of scientists with talent and commitment adequate to ensure success, and technological readiness.

CONTEXT FOR PRIORITIES IN FEDERALLY SUPPORTED RESEARCH

Priorities for federally funded space science, indeed for federally sponsored scientific research generally, should be considered in the light of national goals. Some scientists argue that expansion of our understanding of the physical and biological world will inevitably produce a future flow of societal benefits and that such benefits will justify federal support, but this position does not respond to the current demand for a more direct link between research and national goals. A hierarchy consisting of national goals, strategic endeavors, and specific initiatives can be viewed as the framework in which scientific priorities are determined.

National Goals

The broadest category and the top level of this hierarchy is the set of national goals that express the most basic aims of our society. The following list is not exhaustive, and it is not arranged in order of significance. It is offered to illustrate the general objectives that motivate a wide range of individual and governmental activities. National goals include:

- Defending personal liberty and democratic principles;
- Expanding human knowledge and understanding;
- Promoting and ensuring economic vitality;
- Securing justice for all citizens;
- Providing opportunity for meaningful and rewarding work for all citizens;
- Providing improved education for all citizens; and
- Preserving the environment.

A typical scientific initiative may contribute to a few, but probably not all, of such national goals.

The relative emphasis on such goals and the language in which they are stated changes with time and in response to particular historical contingencies such as external threats to national security, the state of the national economy, the need to mitigate social distress, or opportunities for technological development. Despite sometimes changing emphasis and meaning, these goals are pursued on very long time scales.

Strategic Endeavors

The next level of the hierarchy is that of strategic endeavors composed of activities contributing to the achievement of national goals. Strategic endeavors are devised to address broad societal problems and needs. Examples include the so-called war against cancer, the study of global change induced by human activities, the development of an enhanced computer and information infrastructure, and the scientific exploration of the solar system. Strategic endeavors often involve several federal agencies and a wide range of institutional and individual participants. In recent years, some have been explicitly identified in the President's budget or institutionalized as national research programs. Five areas of particular interest that were identified by the Federal Coordinating Council on Science, Engineering, and Technology[17] are high-performance computing, advanced materials and processing, biotechnology, global change, and mathematics and science education. Two of these, high-performance computing (PL 102-94) and global change

(PL 101-606), are authorized by statute. Strategic endeavors are pursued over periods of years and often decades.

Specific Initiatives and Activities

At the third level of the hierarchy are the specific initiatives or activities through which the aims of strategic endeavors are actually achieved. For space science, these include focused research programs, space missions, technology development programs, and development of new research facilities. The conceptualization, development, and implementation of an initiative may take years to a decade or so. For this reason, the task group recognizes two categories of specific initiatives, namely, programmatic activities, and conceptual or developmental efforts.

In space research, programmatic activities include ongoing research and design, construction, and flight of spacecraft and the exploitation of data from such flights. Examples of present programmatic activities include the Advanced X-ray Astrophysics Facility (AXAF) and the Earth Observing System (EOS).

In contrast, conceptual efforts develop and propose new ideas and new approaches for addressing scientific questions. They also examine possible uses of advanced technology to obtain scientific information. In space research, they often explore mission concepts, refining them until they evolve into proposals for programmatic activities. Developmental or conceptual efforts are typified by proposals for an astronomical facility on the moon, Mars penetrators, or a constellation of geosynchronous satellites for continuing surveillance of the Earth and its atmosphere.

Scientific Priorities

The relative emphasis to be given individual national goals is determined by the national political process. At the next level, the identification of strategic endeavors and the assignment of priorities to them is accomplished by the federal government, taking account of scientific and technological progress and opportunities, and shaped to some degree by the advice of the scientific community through various advisory mechanisms. At the third level, priorities for programmatic activities are determined as part of the process of preparing the federal budget and are often shaped in important ways by advisory committees. However, in most areas, including space research, there is no formal process for obtaining advice from the scientific community on relative priorities of developmental efforts.

In the case of programmatic activities or missions, disciplines do set forth long-term internal strategies that identify or rank developmental efforts, often through consensus building by advisory committees. As emphasized throughout this report, however, advisory committees only recommend priorities; priorities are actually set and programs approved and funded via administration and then legislative action. In a stable or expanding budget environment, recommendations by discipline practitioners may be given great weight. In an environment of contracting resources, however, external forces on program definition are increasingly influential, first outside the agency, and then, as the pressure increases, on senior executives within the agency as well. As a result, priority outcomes may diverge from the originating advisory groups' expectations in scope and scale, but should remain faithful in overall direction.

TAKING UP THE CHALLENGE:
A METHOD AND AN EXPERIMENT

Following its reorganization in 1988-1989, the Space Studies Board created the Task Group on Priorities in Space Research and charged it to investigate the feasibility and utility of recommending priorities for space science in the strategic domain extending between 5 and 15 years from the present. The presumption was that planning within the five-year time horizon was the province of agency programmatic scheduling and decision making, and that focusing on the longer-term domain was more consistent with the strategic planning and assessment charter of the Board. The task group was further charged, in the event that long-term priority setting was determined to be a promising concept, to develop an operational methodology that the Board could use.

In early 1992, the task group released its first report, *Priorities in Space Research: Opportunities and Imperatives*, which found that long-term strategic planning was not only desirable but also necessary to promote a rational and optimized space research program. After release of this report, the task group proceeded to construct a model process by which the recommended priority setting could be carried out.

The task group developed a first version of a two-part priority-setting instrument, which consisted of an advocacy questionnaire and an evaluation aide, and performed a trial of the instrument during a retreat at Woods Hole, Massachusetts, in the summer of 1992. As a result of this experiment, the instrument was improved and simplified. In February 1993, the revised tool set was used by the full Board in carrying out a larger and more realistic trial. The Board was well suited as a tester for the instrument for several reasons: if the instrument proved successful, the Board could use it or an operational version to prioritize proposed long-term research initiatives as part of its regular advisory program. Further, because of the Board's multidisciplinary composition, its reaction to the task group's method would be a good predictor of the method's likely acceptance in the broader research communities were it to be put into operational use.

After the February 1993 Board trial, the task group prepared a final report on the methodology and its two experimental applications. The Board discussed the report and the experiments three times in 1993, at its February, June, and November meetings, and in the end decided not to adopt the recommended priority-setting procedure for routine use. This report presents the motivation, nature, and application of the task group's priority-setting instrument and method, and summarizes the Board's reaction to it.

Chapter 2 lays out the general features of a priority-setting methodology, Chapter 3 and Appendixes A through D describe the tools developed by the task group, Chapter 4 recounts the two experimental trials of the tools, Chapter 5 discusses the Board's reaction to the instruments and the task group's early final report drafts, and Chapter 6 summarizes the findings of the whole activity.

NOTES

1. Space Studies Board, National Research Council, *Setting Priorities for Space Research: Opportunities and Imperatives*, Task Group on Priorities in Space Research, National Academy Press, Washington, D.C., 1992.

2. National Aeronautics and Space Administration, Office of Space Science and Applications, *Strategic Plan—1988, 1989, 1990, 1991*, Washington, D.C.

3. National Aeronautics and Space Administration, *The Crisis in Space and Earth Sciences—A Time for a New Commitment*, Space and Earth Science Advisory Council, Washington, D.C., 1986.

4. Dutton, J.A., and Lawson Crowe, "Setting Priorities Among Scientific Initiatives," *American Scientist* 76, 599-603, 1988.

5. Federal Coordinating Council on Science, Engineering, and Technology (FCCSET), *Our Changing Planet: The FY 1993 U.S. Global Change Research Program, A Supplement to the President's Fiscal Year 1993 Budget*, Committee on Earth and Environmental Sciences, Office of Science and Technology Policy, Washington, D.C., 1993.

6. National Research Council (NRC), *The Decade of Discovery in Astronomy and Astrophysics*, Astronomy and Astrophysics Survey Committee, National Academy Press, Washington, D.C., 1991.

7. Institute of Medicine (IOM), *Setting Priorities for Health Technology Assessment—A Model Process*, Committee on Priorities for Assessment and Reassessment of Health Care Technologies, National Academy Press, Washington, D.C., 1992.

8. Hatry, Harry P., Annie P. Millar, and James H. Evans, *Guide to Setting Priorities for Capital Investment*, The Urban Institute Press, Washington, D.C., 1984.

9. The focus of the SSB inquiry is thus more specific than that addressed in the report *Science, Technology, and the Federal Government: National Goals for a New Era* issued by the NRC Committee on Science, Engineering, and Public Policy (National Academy Press, Washington, D.C., 1993), which sought to describe the implications of the changing "framework within which the U.S. research and development system functions." As national science goals, the report proposed that the United States should be among the world leaders in all major areas of science, that the United States should maintain clear leadership in some major areas of science, that the comparative performance of U.S. research in a major field should be

assessed by independent panels, and that implementation of these goals requires a more coherent federal budgetary procedure.

10. The White House, National Space Policy, November 2, 1989.

11. Advisory Committee on the Future of the U.S. Space Program, Summary and Principal Recommendations of the *Report of the Advisory Committee on the Future of the U.S. Space Program*, U.S. Government Printing Office, Washington, D.C., 1990.

12. Office of Science and Technology Policy, letter to Advisory Committee on the Space Station, April 30, 1993.

13. Space Science Board, National Research Council, *Space Science in the Twenty-First Century: Imperatives for the Decades 1995-2015 (Overview)*, National Academy Press, Washington, D.C., 1988.

14. This assumption is not necessarily incompatible with the civil space program's continuing interest in establishing and maintaining a human presence in space. First, there is much to be learned about how humans respond to the space environment and what measures will make them more effective there. Second, the presence of humans would be scientifically justified when timely and compelling attempts to obtain information and understanding in space require human participation.

15. Brown, George E., "Rational Science, Irrational Reality: A Congressional Perspective on Basic Research and Society," *Science*, 258, 200-201, October 9, 1992.

16. Committee on Science, Space, and Technology, *Report of the Task Force on the Health of Research*, U.S. House of Representatives, 102nd Congress, 2nd Session, Serial L (July), U.S. Government Printing Office, Washington, D.C., 1992.

17. Office of Technology Assessment, *Federally Funded Research: Decisions for a Decade*, U.S. Government Printing Office, Washington, D.C., 1991.

2

The Process of Setting Priorities

Priorities must be determined, implicitly or explicitly, in any situation in which the resources required to pursue a collection of objectives exceed the total available. This situation prevails in the disciplines of space research and in the entire civil space program, as discussed in the task group's first report,[1] and is present in a variety of scientific and other endeavors. This chapter introduces the key issues in designing a process for setting priorities.

CHARACTERISTICS OF AN EFFECTIVE PRIORITY-SETTING PROCESS

The basic assumption in developing a priority-setting process is that a reasoned and structured approach for ordering competing initiatives will yield more and better results than other ways of proceeding. In order to actually produce such a desirable outcome and to ensure that the results are generally accepted, a process for determining priorities should satisfy a number of requirements. It should:[2]

- *Reflect the values of the organizations using it.* The process and recommendations must be consistent with the mission and values of the organization(s) to which the recommendations are directed and must also reflect the values of the persons or groups developing the recommendations. Otherwise both conflicts and confusion are likely. It is not possible to develop a reasoned and convincing answer to the question "What should we do?" unless what is truly important to the parties involved is known.

• *Generate confidence by being sensitive to the political and social context in which it is used.* To be effective, a process for setting priorities should be acceptable to those who are affected by it. The process must be understandable and logically defensible, it must be justifiable and traceable, it must be objective and fair, and it should be open to initial input from all those involved. Moreover, the process must reflect any unavoidable constraints, and their implications must be made clear at the outset.

• *Be efficient.* The process must ensure that important issues are addressed, and conversely, it must ensure that relatively unimportant issues are excluded as quickly and inexpensively as possible. Similarly, in the interests of openness, many candidates must be considered, but those of low priority should be discarded early enough that the process can concentrate on the higher-priority candidates. Finally, the process must be efficient in the sense of being inexpensive (in both time and money) relative to the costs of implementing the initiatives under consideration.

• *Provide a useful product.* To be effective, the product of a priority-setting process must be focused on the needs of the organizations or communities to which the recommendations are directed. If the process does not produce advice that is relevant to the recipients, it will be ignored.

KEY ELEMENTS OF A PRIORITY-SETTING PROCESS

A number of issues must be addressed in developing a process for recommending priorities. Four key elements that must be specified before the process of evaluating and ranking proposals can be started are the set of candidates to be considered, the criteria to be applied, a method for describing the candidates, and a method for evaluating and ranking them. One of the lessons learned in the task group simulations (Chapter 3) is that these four key elements are not mutually independent: a method for describing and evaluating one class of candidates may not be appropriate for another. Care must be taken to ensure that the interrelationships among these key elements are understood before seeking proposals and starting evaluations.

Determining Categories of Candidates

Determining which activities or initiatives are to be considered is a first step in a priority process. In its first report, the task group argued that

> [a]ttempts to set priorities in scientific research should concentrate on specific initiatives or proposals for activities at the margins of ongoing efforts. . . . it is impossible to rank the disciplines of science or space

research in a priority order. It is essential to concentrate on the initiatives produced by the disciplines, not the disciplines themselves.[3]

Initiatives that are similar in scope, complexity, and cost can be compared; when the entire ensemble of initiatives contains subgroups that differ substantially, they must be divided into categories that are relatively homogeneous. In this case, priorities must be determined first among the initiatives within each category and then between the categories, leading to a two-dimensional priority structure. The breadth of effort in some federally funded research areas, such as space research, mandates that the diversity of proposals be anticipated before proposals are solicited. As an example of this diversity, space research includes the efforts of individual investigators in basic research, identifiable research programs involving many investigators but focused on common objectives or themes, and complex and costly projects to create and operate a spacecraft or a research facility.

Cost is clearly a criterion for categorizing initiatives; space research initiatives vary from research programs with annual costs of tens of millions of dollars to proposals for complex missions with direct annual costs exceeding hundreds of millions of dollars. Moreover, there is a U-shaped relationship between cost of an initiative and the scale of, or distance to, the object of interest. Scientific study of processes with very small scales (as with high-energy accelerators) and very large scales (such as study of an entire planet) are vastly more expensive than the study of more readily accessible processes on a human scale. Thus, a collection of experiments aimed at revealing the consequences of microgravity for physical or biological processes is fundamentally different from a flight project to observe a planet.

Maturity is another critical attribute, regardless of the nature of the scientific discipline or the scale of the objects of interest. Conceptual efforts cannot be compared to initiatives that have matured sufficiently so that detailed, well-defined plans for implementation alternatives are available.

Specifying Criteria

Explicit specification of the criteria that will be used to judge proposals is a critical feature that distinguishes a formal, objective priority process from subjective or informal processes in which the criteria are not clearly understood by either the proposers or the evaluators. The task group believes that the following four criteria are appropriate for judging scientific initiatives.

Scientific Merit

The scientific merit and benefits of an initiative are primary considerations in setting priorities for scientific initiatives. Scientific merit must be judged on the basis of the scientific questions to be addressed, the impact of the answers to these questions on the proposing discipline and on the broader scientific enterprise, and the ability of the initiative to provide meaningful answers to the motivating questions. In its first report, the task group argued that scientific merit is judged by assessing whether the initiative, if successfully executed, will provide new knowledge in the form of discoveries or deeper understanding of physical or biological phenomena, structure, or evolution. One measure of the relative scientific merit of an initiative is the degree to which it "contributes most heavily to and illuminates most brightly its neighboring scientific disciplines."[4] Moreover, merit must be judged relative to phases of science, including (1) exploration or discovery, (2) reconnaissance, observation, or experiment, and (3) theory, modeling, and simulation.[5] The greatest scientific benefit in a mature discipline may come from theory; in a new discipline, from exploration. Conversely, a mature discipline may change dramatically because of a new discovery, or an emerging discipline may leap forward because of possibilities revealed by a new theory.

Contributions to National Goals

With the increasing emphasis on contributions by federally funded research to national goals, it is ever more important that scientists be able to describe and justify the benefits of initiatives to the public and their representatives. If an initiative does not contribute to national goals, there is little justification for supporting it with public funds. Most scientific initiatives contribute primarily to the goal of expanding human knowledge. But many contribute elsewhere as well, including economic vitality through applications, technological progress, and improved decision making; improved education; and preserving the environment. While contributions to enhanced understanding may be emphasized by scientists, the task group believes that scientists will benefit if they analyze the full range of potential benefits and are mindful of their importance to others.

Cost

Anticipated costs are a key criterion since it is resource limitations that force priority setting in the first place. Estimating the costs of some space research initiatives is difficult because immediate project costs may be only part of a total cost that also includes launch vehicles, communications systems, and data systems. Many of these costs have been transparent in the past in space research because it has been NASA practice to provide the necessary infrastructure (including launch services) for space research from operational rather than science budgets. Also, high-level policy shifts in the past have on occasion forced individual payloads to be reassigned to a different launch vehicle, driving up costs unpredictably. Another problem is how to estimate costs for extended missions; these extensions are discretionary to agency management, and decisions about them are made after an initial period of operation based on factors such as the relative productivity of the mission once under way, competition from other funding needs, and external budget and policy influences.

Estimating costs in categories with relatively broad ranges is probably advisable because of the uncertainties of determining costs for conceptual initiatives or other future projects. Careful comparison with similar projects whose costs are actually known may give a sufficiently accurate estimate. In any case, the priority-setting process must specify a method for estimating costs and use a consistent approach within each category of initiatives.

Likelihood of Success

Many obstacles may prevent successful completion of space research initiatives, ranging from inadequate management to catastrophic failure of research instruments or equipment. It is important to assess the risks faced by an initiative and to estimate the likelihood of success. Given two initiatives equal in all other criteria, investment in the less risky one would be preferable. In contrast, an initiative's potential for yielding especially significant benefits might justify accepting a relatively small likelihood of success. Estimating the likelihood of success involves considering technological requirements and readiness, requirements for data and information systems, and the vigor, experience, and commitment of the associated scientists, engineers, and managers.

Describing Candidate Initiatives—The Advocacy Statement

Once the categories of candidates and the criteria are determined, it is necessary to determine how the initiatives are to be described for evaluation and ranking. It is assumed that advocates proposing an initiative will provide a description in a specified format. Earlier experience[6] has demonstrated the critical importance of thorough, consistent, and written documentation describing initiatives relative to established criteria.

The task group developed a template for describing initiatives whose total costs are expected to equal or exceed several hundred million dollars and tested it in two simulations. The simulation exercises demonstrated that the template was inappropriate for research programs composed of a variety of efforts and experiments. This finding motivated the development of a simpler template consisting of a series of questions to guide the development of an advocacy statement. Further discussion and the final template are given in Chapter 3 and Appendix A, respectively.

Evaluating Initiatives

Approaches for evaluating and prioritizing initiatives range from informal, subjective schemes to formal, quantitative methods. In the simplest scheme, each member of a group evaluating a category of initiatives reads the advocacy statements and prepares a ranked list. These lists are then combined by assigning a numerical value to each rank and determining the average rank of each initiative.

A more structured approach guides the evaluators to fully assess the advocacy statement in terms of stated criteria. Evaluation forms can be constructed that mirror the advocacy statement and ask the evaluators to respond to qualitative assessment questions or to complete the same quantitative scorecards already filled in by the advocates. An abbreviated, qualitative evaluation form is presented in Appendix B. Alternatively, evaluators can be asked to provide summary scores, perhaps one relative to each of the major criteria discussed above. Appendix C presents a quantitative evaluation form to accompany the detailed template in Appendix A.

Fully quantitative evaluation schemes have been proposed by other groups[7] and are described in Appendix D, but were not tested by the task group.

A SCHEMATIC SEQUENCE FOR SETTING PRIORITIES

The process of developing priorities for space research involves several separate groups of people, including the organization that supports the research, the group charged with the responsibility for recommending priorities, and the scientists or teams that are developing initiatives and will be the advocates of the initiatives that will be considered.

Formal procedures to be followed by the group recommending priorities and developing an agenda can be constructed as a series of distinct steps for selecting and ranking the initiatives. As a guide for developing a formal priority-setting process, the task group has divided the procedure into 11 separate activities. In the discussion, it is assumed that responsibility for recommending priorities resides in a "Committee."

Design the Key Elements

1. *Specify the candidates and categories.* The first (and often difficult) step is to specify exactly what will be considered and what will be excluded from the process. If the collection of eligible candidates is sufficiently heterogeneous, then appropriate categories must be specified.
2. *Specify the criteria.* The criteria that will be addressed in preparing advocacy statements and in carrying out the evaluation must be stated explicitly and clearly. The task group believes that broad criteria, such as those presented above, are useful as categories within which to assemble more specific criteria (as illustrated by the template in Appendix A).
3. *Formulate proposal templates.* The evaluation process will be greatly simplified if the advocates prepare statements in accordance with formats or templates specified for all proposals within each category of candidates.
4. *Specify the evaluation procedure.* Evaluation schemes can range from subjective and qualitative approaches to formal, quantitative procedures. If quantitative procedures are to be used, then the weights to be applied to each of the criteria must also be specified.

To ensure that all relevant issues are addressed thoroughly and openly, the Committee may wish to communicate the results of these four steps to potential proposers and ask for comments and suggestions.

Obtain Proposals or Advocacy Statements

5. *Obtain proposals.* Once the priority process is specified in final form, advocacy statements may be sought from potential proposers.
6. *Verify category assignments.* An initial screening verifies that proposals submitted for consideration in a specific category meet the requirements for that category and are prepared in accordance with specifications.

Evaluate and Rank Candidates

The evaluation and ranking of initiatives can be performed by the entire Committee or by subcommittees given responsibility for single categories.

7. *Evaluate and rank candidates within each category.* Each evaluator (a member of the Committee or a subcommittee) should have a copy of each proposal within the category being considered and an evaluation form to record the results of examining the proposals. Once individual evaluations are completed, the summary evaluation and ranking can proceed according to the already-specified procedure. At this stage, tests of the sensi-

tivity of the results to weights or composition of the evaluation groups may be useful to establish confidence in the results or to identify issues that must be addressed before proceeding.

8. *Rank categories or assign relative fractions of effort.* There are two possibilities for completing the priority structure after initiatives are ranked within categories. In the first, the Committee develops a procedure for ranking the categories, thus producing a matrix of priorities. In the second, each category is assigned a fraction of the total effort or resources; this assignment might be made in advance by the sponsoring organization.

Final Steps

The priority rankings within and between categories are the raw material that allow the Committee to prepare a proposed research agenda.

9. *Prepare the agenda and summary document.* Given external constraints such as annual funding available, the relative rankings can be converted into an agenda that, if not a schedule of year-by-year activities, at least recommends an order in which the initiatives are to be pursued (or parallel orders in the case of several categories). The agenda should be stated explicitly and justified as fully as possible in a preliminary summary document.

10. *Perform a "sanity test."* In this very important step the group preparing the priority recommendations pauses and reviews its work, asking, does this all make sense? Are there any issues left unresolved? Is there a potential for surprises or unexpected adverse consequences? Any attempt to use this step to seek special favor must be resisted, however. The real purpose is to instill confidence in the Committee that its recommendations can be defended.

11. *Publish and advocate the results.* Finally, a report detailing the priority-setting process, the prioritized recommendations or recommended agenda, and a discussion of the implications is published. When the results affect substantial research communities, members of the Committee may be called on to spend considerable time presenting them in a variety of forums. This explanation and advocacy of the results may be critical to ensuring that the recommendations are accepted by the communities involved.

NOTES

1. Space Studies Board, National Research Council, *Setting Priorities for Space Research: Opportunities and Imperatives*, Task Group on Priorities in Space Research, National Academy Press, Washington, D.C., 1992.
2. This discussion is based on the Institute of Medicine's *Setting Priorities for Health*

Technology Assessment—A Model Process, Committee on Priorities for Assessment and Reassessment of Health Care Technologies, National Academy Press, Washington, D.C., 1992.

3. Space Studies Board, National Research Council, *Setting Priorities in Space Research: Opportunities and Imperatives*, Task Group on Priorities in Space Research, National Academy Press, Washington, D.C., 1992, pp. 54-55.

4. Weinberg, A.M., "Criteria for Science Choice," *Minerva* 1, 159-171, 1963.

5. Dutton, J.A., and Lawson Crowe, "Setting Priorities Among Scientific Initiatives," *American Scientist* 76, 599-603, 1988.

6. See Notes 3-5 above.

7. Institute of Medicine, *Setting Priorities for Health Technology Assessment*, 1992, op. cit.

3

The Tools: The Advocacy Statement and Evaluation Form

The task group developed two main tools for use in setting priorities. The first is a template for a detailed advocacy statement in which proposers of an initiative describe how it meets the criteria proposed in Chapter 2. The second is an evaluation form, mirroring the advocacy statement, that aids evaluators in assessing the initiatives. Two versions of the evaluation form were developed, a *qualitative* form and a *quantitative* form. The task group tested the detailed advocacy statement and the evaluation form in two simulations. In the first, the entire process of writing advocacy statements, evaluating them, and determining rankings was carried out over a three-day period by a group composed of members of the task group and representatives of the Space Studies Board disciplinary committees. As a result of this first test, both the advocacy statement and the quantitative evaluation form were simplified; the qualitative form was dropped as discussed in Chapter 4. The resulting revised advocacy statement and evaluation forms are described in this section.

THE ADVOCACY STATEMENT

The advocacy statement provides information on how a proposed initiative satisfies the stated criteria. Based on the criteria proposed in Chapter 2, important questions are as follows:

- What are the scientific issues and questions motivating this initiative?

- What are the scientific strategies and proposed research activities for addressing these issues?
- What scientific benefits may be expected?
- What contributions to achieving national goals are expected?
- What capabilities and facilities exist now for supporting the initiative? What capabilities and facilities must be developed?
- What mechanisms and budgets exist or are proposed for supporting and managing this initiative? What will the total costs be?
- What risks imperil this initiative and how can they be mitigated?
- What is the likelihood that the proposed initiative will achieve its scientific objectives?

This set of questions could serve as a template for describing focused research programs within a discipline, those with annual expenditures in the tens of millions of dollars.

For larger efforts, typified by the long-term development of a new flight project, a more detailed template and advocacy statement are appropriate. The template for the detailed advocacy statement proposed by the task group is presented in Appendix A. It addresses the criteria of scientific merit, contribution to national goals, cost, and likelihood of success both qualitatively and quantitatively. At various points, the template asks the proposers to summarize their written accounts using quantitative assessments in major areas. (The same scorecards appear in the detailed evaluation form presented in Appendix C, thus guiding the evaluators to consider the same issues for each proposed initiative.) Although the scorecards ask for numerical evaluation on a simple scale, they are not intended as components of an overall score. Rather, they are meant to stimulate both advocates and evaluators to think carefully about major components of the proposals.

The advocacy statement format defined by the task group is structured into the following principal components.

Context

The template asks the advocate to describe the complete case for the initiative in a concise statement that illuminates the key contributions and benefits. The task group trials demonstrated that effective introductory statements can be framed in just a few pages. Next, a description of the background, history, and context of the initiative is requested. This information helps place the initiative properly in a broad scientific and national setting, and sets the stage for more specific answers later in the statement. In the trials, the introductory statements and the information on history and context proved useful in understanding the initiatives.

Scientific Merit

The scientific merit and benefits of an initiative must be judged on an array of attributes. These include the scientific questions addressed by the initiative, the impact of the answers to these questions on a discipline and on the broader scientific enterprise, and the ability of the initiative to provide meaningful answers to the motivating questions. In its first report, the task group argued that the fundamental question in judging scientific merit is whether the initiative, if successfully executed, will provide new knowledge in the form of new discoveries or deeper understanding of physical or biological phenomena, structure, or evolution.

All of these aspects of scientific merit are explored in a structured manner in the template for the advocacy statement. The template requests explicit statements of the relevant scientific questions and an analysis of how they relate to the discipline and other fields. The advocates are asked to give a quantitative assessment of the relative importance of the scientific questions addressed by the initiative. They are asked to describe any unique features or special motivations of the initiative, and finally, to provide an assessment of the overall scientific significance of the proposal.

Contribution to National Goals

There is an increasing demand for explicit specification of the benefits to society to be expected from federally funded research programs. The template for the advocacy statement emphasizes the three broad public objectives of promoting human welfare, stimulating economic vitality, and enhancing national security, and asks how the initiative serves these objectives. In addition, the template identifies seven specific ways, ranging from education to commercial space activities, in which an initiative might serve larger national goals.

It is anticipated that most space research initiatives will contribute to the achievement of only a limited set of national goals, but it is important in the present political context to identify those that will be served and to provide a compelling argument about the contribution that can be expected.

Programmatic Aspects

The programmatic readiness of major space research initiatives often dictates whether they are put forward for immediate implementation or delayed for further planning and maturation. The template addresses a number of programmatic features of initiatives, including costs, identification and mitigation of risks, and the consequences of failure. There is a particular emphasis on the availability of human resources to accomplish the objectives of the initiative and on planning for data analysis and for subsequent supporting research.

The information requested on plans for data analysis and research addresses the conceptual and implementation aspects of converting observations or other results of an initiative into enhanced understanding, and thus amplifies the description under the section "Scientific Merit" of how the initiative will help resolve the key scientific questions. The task group asserts that the production of new information and understanding is the unifying goal of all space research, and therefore gives it high priority in the template.

Summary

The summary asks proposers to provide their evaluation of the relative value of the scientific and social benefits of the initiative, as well as an assessment of the benefit-to-cost ratio and the likelihood of actually achieving the anticipated benefits.

THE EVALUATION FORM

To arrive at recommendations for relative priority, a set of initiatives must be evaluated and ranked relative to one another. Thus, the group responsible for recommending priorities must develop a collective evaluation of the initiatives as described in the advocacy statements. The task group considered two approaches, a simple qualitative scheme and a detailed semiquantitative scheme, and developed evaluation forms for both.

The simpler, *qualitative* scheme asks evaluators to arrange the initiatives in a simple ranked list. The consensus rank is then the average of the individual rankings. The form for the qualitative approach is provided in Appendix B.

The detailed scheme was designed to assist evaluators to consider fully how well each initiative met the stated criteria by closely following the advocacy statement. The detailed evaluation form asks evaluators to respond to qualitative questions and to provide their own numerical rankings on the same quantitative scorecards previously completed by the advocates. In the trials conducted by the task group, the evaluators were also asked to provide their own recommendations for relative weights for scientific merit, contribution to national goals, and programmatic aspects. One advantage of such an approach is that a consensus set of values for the weights can be determined and applied to all individual rankings to arrive at a consensus rank. A form for the detailed, *quantitative* scheme is provided in Appendix C.

4

Testing the Tools and Methodology

Chapter 2 outlines the characteristics and components of an effective priority-setting scheme. During its work, the task group constructed the standard format for initiative advocacy and several versions of the evaluation instrument discussed in Chapter 3. These were applied in two trials of the resulting priority-setting methodology using model initiatives. This section discusses the two evaluation procedures and the results of applying them to a hypothetical suite of initiatives in simulations of the priority-setting process.

TESTING THE PROPOSED PRIORITY-SETTING PROCESS

The task group tested proposed formats for the advocacy statement and evaluation forms in two trial simulations. In the first trial, the entire process of writing advocacy statements, evaluating them, and determining rankings was carried out over a three-day period at Woods Hole in July 1992. Subcommittees formed of representatives of Space Studies Board disciplinary committees and task group members developed plans for hypothetical initiatives and completed advocacy statements for them. A group of 12 evaluators (some task group members and some representatives from the various Board disciplines) then simulated the activities of a committee responsible for recommending priorities. Each participant evaluated each initiative with an initial evaluation form and prepared individual ranked lists. As a result of this test, the advocacy statement and the evaluation form were both simplified. In the second trial, 15 members of the Space Studies Board evaluated initiatives using the revised advocacy statement and evaluation forms during a two-day period.

The task group decided before the trials that only hypothetical initiatives would be considered, for two reasons. The first was the belief that it was essential to avoid any inadvertent interference with any existing, real priority-setting processes. The second was the desire to test the process on a broad range of initiatives that were still in the conceptual phase, as would be the case for initiatives derived from Board discipline research strategies. As a consequence of their hypothetical nature, some of the proposed initiatives did not possess the intellectual or technological maturity that would be expected of real proposals.

Five hypothetical initiatives were considered in the first simulation: two space research missions, an application that combined space observations with data management, and two research initiatives with a space component. Four hypothetical initiatives were considered in the second simulation: two space research missions, an application that combined space operations with data management, and one research initiative with a space component. Three of these initiatives were the same as ones used in the first simulation. Although largely invented for the study, the initiatives were all aimed at real scientific issues. They varied significantly in content, costs, readiness, and the relative importance of scientific merit and social benefits as justifications. These initiatives are designated here by the letters A through E for the first simulation and F through I for the second simulation (with no correspondence to the hypothetical initiatives given above in this paragraph).

DETAILED RESULTS OF THE TRIALS

How the Rankings Varied with Different Summary Criteria

Tables 3.1 and 3.2, corresponding to the first and second trials, respectively, give the relative ranking of the initiatives on the basis of four summary measures from the quantitative evaluation. *Program benefits (unadjusted)* refers to the total score awarded for the benefits expected to accrue from an initiative, averaged over all evaluators. *Program benefits (adjusted)* refers to a similar average score for each initiative, but excluding respondents with a disciplinary interest in that initiative in order to correct for possible bias. The *benefit-to-cost ratio* summarizes evaluators' views of the relative ranking of initiatives when benefits are compared to costs, and is presented also both unadjusted and adjusted for bias.

The results for both trials show that the two measures, program benefits and the ratio of benefits to cost, result in different rankings for the proposed

TABLE 3.1 First Trial: Quantitative Ranking of the Initiatives

Rank	Program Benefits (Unadjusted)	Program Benefits (Adjusted)	Benefit-to-Cost Ratio (Unadjusted)	Benefit-to-Cost Ratio (Adjusted)
1	A	A	A	D
2	E	E	D	B
3	B	B	B	A
4	D	D	E	E
5	C	C	C	C

TABLE 3.2 Second Trial: Quantitative Ranking of the Initiatives

Rank	Program Benefits (Unadjusted)	Program Benefits (Adjusted)	Benefit-to-Cost Ratio (Unadjusted)	Benefit-to-Cost Ratio (Adjusted)
1	H	H	H	H
2	F	F	I	G
3	G	G	G	F
4	I	I	F	I

initiatives. This difference between these results is important, as it could be argued that the benefit-to-cost ratio is the most appropriate measure for judging initiatives in a world where both benefit and cost are important decision criteria.

A comparison of unadjusted and adjusted measures to examine potential respondent bias toward respondents' own disciplines reveals that such a bias was absent from the scoring of program benefits in both simulations, but present in the scoring of benefit-to-cost ratios. Examination of the evaluations revealed that evaluators gave top total summary scores to initiatives associated with their own disciplines in about half the cases (in the first trial, 56 percent of the time, and in the second, 45 percent of the time). In the trials reported here, the important benefit-to-cost ratio appeared to compensate for potential respondent bias.

Comparison of Quantitative and Qualitative Evaluations

In the first trial, evaluators completed both quantitative and qualitative evaluations (see Appendix B for the qualitative form). Table 3.3 lists these scores for the first trial. The rankings are not consistent. If, as suggested

TABLE 3.3 First Trial: Comparison of Quantitative and Qualitative Rankings

Rank	Quantitative Ranking			Qualitative Ranking[a]
	Program Benefits[a]	Benefit-to-Cost Ratio (Unadjusted)	Benefit-to-Cost Ratio (Adjusted)	
1	A	A	D	A
2	E	D	B	B
3	B	B	A	D
4	D	E	E	E
5	C	C	C	C

[a]Adjusted and unadjusted rankings are identical.

above, the benefit-to-cost measure is the most appropriate measure, then this finding argues against a solely qualitative evaluation. For this reason, only the quantitative evaluation was used in the second trial, carried out later by the full Board.

Weighting by Benefit Categories

Evaluators were free to assign relative weights to the main criteria assessed in the advocacy statements; averaged across all respondents and all initiatives, the evaluators clearly emphasized scientific merit. In the first trial the distribution was scientific merit, 60 percent, contribution to national goals, 23 percent, and programmatic readiness, 17 percent. In the second trial, the distribution was scientific merit, 60 percent, and contribution to national goals, 40 percent. Programmatic readiness was eliminated as a summary criterion in the revised evaluation form used in the second trial.

The weights selected by different populations of evaluators—for example, policymakers compared to scientists—might be expected to differ substantially. Thus, asking evaluators to assign weights explicitly to the main criteria may be important in clarifying their perspectives in interpreting the value of the initiatives. The basis for decisions may therefore be improved as the relative importance of the criteria to the different groups involved becomes clear.

Additional Observations

Several additional inferences can be drawn from the numerical summaries of the evaluations.

Estimated Costs

Cost estimates for the initiatives given in the advocacy statement (averaged across all evaluators and all initiatives) were judged reasonable in 22 percent and 10 percent of the cases (where the first figure applies to the first trial and the second figure to the second trial), within a factor of 1.5 in some 18 percent and 17 percent of the cases, underestimated by between a factor of 2 and an order of magnitude in some 47 percent and 25 percent of the cases, and underestimated by more than an order of magnitude in 12 percent and 48 percent of the cases. This outcome should focus advocates' and evaluators' attention on the need to accurately estimate project costs and judge their plausibility; specifically, the apparent tendency for advocates to underestimate costs, as judged by evaluators, has important implications for the ranking of proposals.

Expected Benefits

In the first trial, on average the evaluators judged there to be a 60 percent chance of success that a proposed initiative's claimed benefits would be realized, with a range of 90 to 45 percent. The highest-scoring proposal on the basis of benefits and the unadjusted benefit-to-cost ratio was judged to have a 66 percent likelihood of achieving the claimed benefits. The highest-scoring proposal on the basis of the adjusted benefit-to-cost ratio was judged to have a 63 percent likelihood of success.

In the second trial, evaluators were asked to rate the likelihood of each proposed initiative's success in attaining claimed scientific objectives and social benefits within the proposed time frame and costs, based on a comparison with other initiatives with which the evaluator might be familiar. On the basis of scientific objectives, and averaged across all evaluators and initiatives, 75 percent of the responses were that the initiatives were "at least as likely," and in 39 percent of the cases, "very likely" or "as likely as the top 20 percent" of initiatives in general, to achieve the stated benefits. On the basis of social benefits, and averaged across all evaluators and initiatives, 95 percent of the responses found that the initiatives were "at least as likely," and in 25 percent of the cases, "very likely" or "as likely as the top 20 percent" of initiatives in general, to achieve the stated benefits.

Significance of Benefits

In the first trial, across all proposed initiatives and evaluators, the assessment of the significance of benefits ranged from 9 (very significant, direct, demonstrable impact) to 4.5 (some direct, demonstrable impact). In the second trial these scores ranged from 10 (same definition as 9) to 2 (indirect or little impact). In the first trial, when multiplied by the scores given for "likelihood of realizing benefits," the highest and lowest scores became 5.7 and 3 (this likelihood was reworded in the second simulation as described in the preceding paragraph and cannot be multiplied to give a probability). Still, it appears that both measures—significance of benefits and likelihood of realizing benefits—are important.

Also in the first simulation, evaluators estimated at 65 percent the likelihood of success of the initiatives within the stated time frame and costs, with a range of 50 to 80 percent. For the highest-scoring proposal on the basis of benefits and the unadjusted benefit-to-cost ratio, half of the evaluators judged the likelihood of success to be 50 percent. For the highest-scoring proposal on the basis of the adjusted benefit-to-cost ratio, the likelihood of success was estimated at 60 percent. The second-highest-scoring proposal on the basis of all measures received a score of 80 percent. These results suggest that the evaluators were willing to accept some risk and that a portfolio of projects—in this case, the two top-scoring proposals—could hedge this risk.

Significance of Initiatives

Evaluators in the first trial were asked to judge the significance of each initiative to the proposing discipline, to space research, and to science generally. On average across all evaluators and proposals, the initiatives were viewed as having decreasing significance across the range from the proposing discipline to science generally in 51 percent of the cases and as being of equal significance in 12 percent of the cases. In the second trial, evaluators were asked to judge each proposed initiative's significance to the proposing discipline and to science generally. On average across all evaluators and proposals, equal significance was accorded in 18 percent of the cases and greater importance to a specific discipline than to science generally was accorded in 82 percent of the cases. As with the weights assigned to benefits, these scores might also be expected to differ among different types of reviewers (e.g., policymakers or space scientists).

GENERAL CONCLUSIONS FROM THE TRIALS

The main conclusions reached by the task group as a result of the tests of the advocacy statement and evaluation forms are as follows:

- The evaluation methodology and instruments were sufficiently general to be applicable to a wide variety of space research initiatives, and offered a useful basis for evaluating and comparing these initiatives.
- The methodology and instruments were found by evaluators to be applicable to evaluations of the hypothetical initiatives.
- Initiatives must be grouped into relatively homogeneous categories and, depending on their scope and overall complexity and cost, evaluated by different procedures. While the methodology seemed to perform well on the space research initiatives, it was too complex for the focused research initiatives. Various approaches to grouping candidates are discussed in Chapter 2.
- The qualitative and quantitative evaluations were both attempted in one of the simulations; overall, they produced different rankings.
- The evaluators' assessments of "benefits" versus "benefit-to-cost ratios" produced different rankings of proposed initiatives in the simulations. The benefit-to-cost ratio may be the more appropriate measure for priority ranking; thus, this ratio should be favored in future approaches to evaluation.
- "Respondent bias" was tested for and seemed to appear in some results of the two trials. Summary rankings were tested for bias by an adjustment that averaged rankings for each initiative over all evaluators except those associated with the research discipline of that initiative. Little bias was found in ranking program benefits, but bias did seem to appear in ranking benefit-to-cost ratios. Evaluators gave top overall ranking to initiatives associated with their own disciplines in about half the cases. Evaluation schemes thus need to take the potential for such bias into account.
- The evaluators assigned an average weight of 60 percent to scientific merit in both trials. In the first trial, the average weight assigned for contribution to national goals was 23 percent and for programmatic readiness was 17 percent; in the second trial no weight was permitted for programmatic readiness, and 40 percent was assigned on average to contributions to national goals.
- Evaluators generally believed that the lifetime costs of the hypothetical initiatives were seriously underestimated, in some cases by as much as an order of magnitude.
- On average, the evaluators judged that there was a medium to high probability for success that the claimed benefits would be realized and that the hypothetical initiatives would succeed.

- The significance of a hypothetical initiative to its associated discipline was generally judged to be higher than its significance to all of science.
- Although the trials helped streamline the task group's forms and procedures and provided useful information, they suffered from a lack of realism.

As a result of these two experiments, the task group concluded that the formats for the advocacy statement and evaluation form were sufficiently mature and validated to support a recommendation that the Board initiate a program of developing priorities for space research using this methodology. This conclusion was reflected in an initial draft of the task group's final report, presented to the Board for approval in June 1993. The task group recognized that hypothetical initiatives are not real initiatives and that priority-setting experiments using them, however thoughtful, can only be simulations. In discussing its results with the Board in June 1993, the task group expressed the view that much more would be learned from application of the methodology to real initiative proposals than from additional experiments.

5

Board Assessment of the Trials and Response to Task Group Recommendations

As described in the previous chapter, the developmental priority-setting methodology devised by the task group was applied in two different trials. The first was carried out by a small panel of volunteers at Woods Hole during July 1992, and the second was conducted by members of the full Space Studies Board at its Washington meeting in February 1993. In all, the methodology was discussed by the Board three times in that and subsequent meetings during 1993.

The Board's initial reactions to the task group's priority-setting instruments were expressed during the discussion at the February 1993 meeting following the second trial of the methodology. Subsequently, the task group prepared a final report on the two trials for approval of the Board and recommended acceptance of the methodology at the June 1993 meeting. Thus, the recommendation of the task group was that the Board should apply the proposed methodology to prioritize real initiatives being considered for implementation 5 to 15 years in the future. During the June 1993 meeting, Board discussion on specifics of the draft final report was interspersed with comments on philosophical issues. The Board was unable to reach a consensus on either the method or the report at this meeting, however, and so they were placed on the agenda again for the November 1993 meeting. In this third meeting, attention shifted back to general questions about the validity of the proposed methodology as the Board continued to try to forge a consensus for publication as a formal NRC report.

This chapter presents a synopsis of the principal concerns expressed during these Board meetings about priority setting and the task group's methodology. Overall, Board members endorsed the concept that practicing scientists should play a role in strategic priority setting. Many comments were made in support of the task group's methodology and recommendations; however, deep divisions persisted among the members, as illustrated by the following condensed summary of the three Board meeting discussions.

BOARD MEETING OF FEBRUARY 1993

Following its test of the task group's priority-setting methodology, the Board devoted agenda time at this meeting to discuss the members' response to the exercise. This discussion ranged over many aspects of the trial, from whether or not priority setting was desirable or feasible to suggestions for improvements to the proposed approach. The following major points were distilled from a transcription of the discussion at the meeting.

General Concerns or Reservations

- Scientists should limit themselves to judging scientific merit, and leave societal judgments to politicians.
- Broad assumptions about the space program must be made explicit before priority setting can be meaningful; for example, the question of whether long-duration human spaceflight will be undertaken affects the value of certain microgravity laboratory investigations.
- There was uncertainty about how technology value could be properly evaluated and compared.
- It was unclear how to accurately measure and represent "social benefit" even if it were to be considered.
- There is a need to define "education" value in proposals; for example, what is the relative value of training graduate students compared to helping with K-12 education?

Concerns About the Methodology or Its Application

- There was disagreement over whether or not initiative developers should make oral presentations; this would improve the information available to evaluators but would confer an advantage on smooth presentations.
- Similarly, if resources available for development of individual proposals differed, there would not be a level playing field.
- Size scales for initiatives are not the same across disciplines, and so there cannot be a level playing field.
- One of the Board's major responsibilities is the preparation of dis-

cipline research strategies; what is the relationship between the "initiatives" in the priority-setting methodology and these "research strategies"?
- It was not clear how to compare relative feasibility of initiatives across discipline boundaries.
- It is necessary to separate the scientific merit of the general objective of an initiative from that of a particular implementation.
- Some members concluded that the methodology should be tried out first in a narrow area, such as a single discipline, while others thought that such a trial would not be worthwhile.

After an extended discussion, the Board decided to wait for the draft final report to be submitted before deciding whether to accept the mandate to adopt the methodology as proposed by the task group. It was expected that the draft report would incorporate results also of the first trial, undertaken at Woods Hole the previous summer. Some members suggested that if a decision were made to proceed with adoption of the test methodology, then a Board group different from the task group would convert the recommendations of the task group into a fully operational "procedure and process."

BOARD MEETING OF JUNE 1993

At its June meeting, the Board reviewed and discussed the draft final report from the task group. The draft stimulated vigorous discussion, some of it related to points raised during the February meeting. Critical comments included the following.

General Concerns or Reservations

- The Board should confine itself to its areas of special competence, that is, science; scientists shouldn't make social judgments.
- The premise that "ultimately we go to space to obtain information and understanding that we can obtain in no other way" (quoted from the task group's first report) is too limited and ignores other reasons to go to space.
- There are "sociological" differences between disciplines, where well-defined and narrow initiatives have an advantage; the way advances are planned and scoped could also lead to prioritization results that would not be accepted by the scientific communities.
- The Board should first make a decision on whether it will undertake cross-disciplinary priority setting at all; the approach proffered by the task group is only one of many possible.

Concerns About the Methodology or Its Application

- Discipline committees might not support the methodology: they develop science strategies, not mission proposals.
- There was some consensus on the Board that the benefit-to-cost ratio is the most valuable evaluation measure; but the ratios are hard to determine, being ratios of poorly determined numbers.
- The methodology proposed does not give enough weight to technology factors.
- Should consideration be given to whether and how the relative degree of a scientific community's enthusiasm for an initiative should be represented in the process?
- It is not clear how to meaningfully compare a cluster of small experiments with a single major mission.
- There was disagreement as to whether scientists from the discipline of a given initiative under consideration should be excluded from the evaluation process; there was evidence that doing so in the trials changed rankings.
- In spite of weighting schemes, experience with similar processes at the National Institutes of Health has shown that they devolved to opinions of those present.
- One member expressed lack of conviction that the methodology had actually been validated by the two trials.
- The methodology needed more real-life testing; its single test by the full Board indicated that it was not a success.

After discussion, the Board was unable to agree on how the report might be revised to be released as a consensus recommendation; it was decided that the task group should reformat its report as an internal report to the Board. It was further proposed that the task group should dissolve, and that the Board should take up the priority-setting issue anew at the upcoming November 1993 meeting. It was later agreed that the report would be revised to remove recommendations and would then be submitted a second time for consideration by the Board for NRC publication as a discussion paper.

BOARD MEETING OF NOVEMBER 1993

At this meeting, the task group submitted a revised version of the final report to the Board for approval and publication. Issues raised during the discussion at this meeting included the following.

General Concerns or Reservations

- There was basic disagreement about whether cross-discipline comparisons could be meaningful.
- It was asserted that it is not possible to prioritize basic research.
- It was questioned whether scientists are the right people to judge "societal benefit"; they have no unique competence.

Concerns About the Methodology or Its Application

- The methodology ignores the central role of individual people, such as principal investigators, when it relies on quantitative measures.
- The methodology does not successfully address differences between areas of science: the revised methodology and instruments recognize the existence of intrinsic differences (e.g., flight projects versus focused research, etc.), but do not assess relative merits across these boundaries.
- The priority-setting process should enforce the separation of primary and ancillary objectives.

After this third discussion, it became apparent that significant and intractable differences remained among the members of the Board. It was decided through a ballot that the final report should be further revised to present the Board conversations as part of the record of the experiment, but with recommendations excised. The present draft was accepted by the Executive Committee of the Board at its meeting in August 1994.

6

Conclusions

There are many reasons for setting priorities in space research or other scientific arenas. Success in space research and the sciences has created more opportunities for exciting initiatives than can be pursued, and so choices must be made. The scientific enterprise, like other national endeavors, should be focused on the most promising opportunities, those that offer the most benefit in improved understanding and in contributing to national objectives.

Public policy should seek to identify initiatives offering the greatest potential for scientific advance and meeting national goals and then pursue them vigorously in accord with a long-term agenda constructed within realistic funding expectations. Such an agenda must be sufficiently focused to provide clear direction, but must be flexible enough to surmount setbacks and take advantage of unexpected opportunities. It must balance the relative likelihood of success against benefits expected from more challenging endeavors. It must provide a balance between disciplines that accounts for different stages of maturity and development. For space science, the agenda should distinguish between flight projects, focused research programs, and the continuing activities in fundamental research, including data analysis, theoretical exploration, and technology development, that are the foundation of the entire effort.

The key question is how the priorities are actually to be determined. The Task Group on Priorities in Space Research has argued that reaching a consensus on the relative priorities for major initiatives would give the space research community and other scientific communities an important opportunity to shape the future of science. By taking account of national goals and the changing context of national imperatives, scientists could contribute significantly to the process through which the critical decisions are eventually made; by recommending an agenda, scientists could help ensure that those decisions are both well informed and enlightened.

To address this opportunity, the task group developed a methodology and instruments for prioritizing research initiatives across discipline boundaries. These instruments consider not only pure scientific merit in ranking initiatives, but also factors such as social benefit and contribution to national goals, risk, and relative benefit and cost. The task group applied this methodology in a series of two tests. The second of these trials was carried out by the full Space Studies Board in February 1993 on sample initiatives developed by its discipline committees. The Board, which includes working researchers representing all fields of space research, provided a convenient and realistic laboratory for exercising the methodology and assessing its efficacy and probable acceptance in the outside research communities.

On three occasions during 1993, the Board discussed at length the methodology and instruments, together with the results of their test application and the task group's report on these tests. These discussions took place at the Board's February meeting, when the Board conducted its trial; at the June meeting, when the Board considered the first draft of the task group's final report; and at its November meeting, when the Board discussed a second version of the final report. During these discussions, members supported the need for scientists to take a greater role in choosing initiatives in space research for execution, and many aspects of the task group's report were favorably noted. The decision the Board faced, however, was a difficult one: in addition to approving the task group's final report, the Board needed to decide whether or not to undertake strategic priority setting across disciplines based on the recommended methodology as a major part of its future responsibilities.

Ultimately, the Board was not able to achieve a consensus on accepting this charge, at least based on the specific tools provided by the task group. After all the discussion, there remained lingering doubt at several levels: whether the Board should do this type of priority setting; whether strictly science-based ranking was possible, and if not, whether the Board had any special expertise in other areas; and finally, a lack of agreement on the effectiveness of the proposed methodology. With respect to the methodology itself, it emerged that the very large initiatives and proposed missions differed in a fundamental way from research efforts involving many smaller projects, suggesting the need for separate and distinct advocacy statements and evaluation schemes. The decision to use only hypothetical initiatives for the trials had led to a markedly heterogeneous set of advocacy statements: one was fairly mature and realistic, while some others were assembled rapidly in an ad hoc fashion. On the other hand, proposals submitted to a real priority-setting process should reflect the maturity that grows from careful and thorough analysis over an extended period of time. It may be that priority setting is like war: simulation, no matter how realistic, is not the same as the real thing because of the stakes. Perhaps only an operational application of a priority-setting method can permit an accurate evaluation.

As a result of these divisions of opinion, the Board was not able to adopt the task group's recommendation that the Board commit to executing its suggested priority-setting program. Instead, the Board recognized the exploratory and analytical value of the work of the task group and elected to ask that the group reformulate its report as an account of the development and use of the methodology and of the Board's assessment of it.

Clearly, setting priorities for science missions and programs will continue to be necessary. Significant benefits may accrue from prioritizing strategic initiatives in advance of formulation of these specific missions and programs. The task group's first report, *Setting Priorities for Space Research: Opportunities and Imperatives* (National Academy Press, Washington, D.C., 1992), makes a strong case, broadly supported by the Board, that scientists should involve themselves in such a process. The subsequent work of the task group, including the cooperation of the Board, demonstrates the difficulty of the priority-setting problem in a strategic and multidisciplinary domain, and illuminates the outcome of one of many possible approaches to solving it. This report provides a record of this effort for the use of others taking up the same challenge.

Appendixes

Appendix A

Advocacy Statement

Template for Advocacy Statement

SUMMARY OF
[NAME OF PROPOSED SPACE RESEARCH INITIATIVE]

Proposers—Affiliations—Addresses

[DATE]

INTRODUCTION: THE PURPOSE OF THIS QUESTIONNAIRE

This questionnaire is designed to assist advocates of initiatives in space science and applications to develop proposals in a consistent format that will facilitate comparative evaluations. It is assumed that initiatives have been considered and recommended as part of the long-term strategies of disciplines in the space and earth sciences and applications. In some cases, initiatives may be potential flight missions or a suite of potential missions. In others, they may be a collection of research experiments to be performed in space. In still others, they may be a proposed program of laboratory research or an element of the research infrastructure. In all cases, it is assumed that the initiatives are not as mature as the mission proposals that are included in the five-year strategic plan of NASA. Rather, these initiatives are candidates for development over a period of five to ten years to sufficient maturity that they might be considered in that plan.

The questionnaire seeks to elicit detailed qualitative and quantitative information about various dimensions of such initiatives. The questions are based on criteria for evaluating scientific initiatives used by NASA's Space and Earth Sciences Advisory Committee and described in its report *The Crisis in Space and Earth Sciences* (1986), and other considerations given by Dutton and Crowe (1988), the OSSA Strategic Plan (1989), and the Phase I report, *Setting Priorities for Space Research: Opportunities and Imperatives*, of the SSB Task Group on Priorities in Space Research.

The questionnaire has four main sections: Scientific Merit, Contribution to National Goals (CTNG), Programmatic Aspects, and Summary. Each main section addresses a specific set of qualitative and quantitative criteria. Please answer all of the questions, using the best available estimates, judgments, and probability assessments. In each section, you will be asked to provide narrative responses that summarize various aspects of the initiative. In addition, you will be asked to summarize the narrative discussions with quantitative assessments that follow from, or are justified by, the narrative descriptions.

Please limit the entire package to about 50 pages, including the tables.

Summary of the [Name] Program or Initiative

> Describe the proposed initiative in a concise statement that illuminates the key contributions and advantages of the proposed program or initiative.

Background and Context of the [Name] Initiative

> Please describe the background and context of this initiative or program, specifying when, how, and why the idea or concept originated and how it matured. Please describe the *scientific context*, and if appropriate, the *national context* (why the initiative is important to the nation), and the *institutional context* (what institutions were involved in developing the proposal and which will be involved in implementing it).

SECTION A. SCIENTIFIC MERIT

This section of the questionnaire addresses scientific objectives and significance of the initiative, its scientific breadth of interest, its potential for providing new discoveries and understanding, and any particularly unique attributes it may have.

1. Scientific Objectives, Significance, and Breadth of Interest

> Please provide concise answers to the following questions.

A.1 What are the key scientific questions addressed by the initiative?

A.2 Why are these questions important to the proposing discipline?

A.3 What impact will the science involved have on other disciplines?

A.4 To what extent is the initiative expected to answer these questions?

Suggested format for response (A.1 - A.4):

Introductory statement about the [name] program and initiative. The key scientific questions (in order of importance) are:

First scientific question.
- Description and explanation
- Scientific significance to the discipline
- Impact on other disciplines
- Data and information to be produced; extent of resolution of the scientific question

Second scientific question.
- Description and explanation
- Scientific significance
- Impact on other disciplines
- Data and information to be produced; extent of resolution of the scientific question

Third scientific question. . . .

TABLE A-1. Evaluation of Scientific Objectives and Significance (Questions A.1 to A.4)

(a) Issues (from A.1)	(b) Significance to Discipline (1 to 10 points)	(c) Broad Scientific Significance (1 to 10 points)	(d) Extent of Current Understanding (0 to 100%)	(e) Extent of Understanding Expected (0 to 100%)
First Issue				
Second Issue				
Third Issue				

Directions:

Column (a): List issues presented in response to question A.1.

Columns (b), (c): Assign points to indicate the relative scientific significance of each issue. Points must total 10 for the column. You may assign all 10 points to one issue.

Columns (d), (e): Indicate the extent of current understanding and the extent of understanding expected from the initiative according to the following conventions:

Percent	Definition
90-100	Complete understanding
80-89	Rapidly developing understanding
50-79	Improved definition of problem; questions identified and clarified
20-49	Preliminary results
< 20	Initial indications

2. Potential for and Likelihood of New Discoveries or Understanding

> Please provide concise answers to the following questions.

A.5 Is there a potential for an important advance in knowledge or understanding, either within a discipline or in areas now separating disciplines? Is there a potential for insight into previously unknown phenomena, processes, or interactions?

A.6 Will the initiative provide powerful new techniques for observing nature? What advances beyond previous measurements can be expected with respect to accuracy, sensitivity, comprehensiveness, and spectral or dynamic range?

A.7 Will the initiative answer fundamental questions or stimulate theoretical understanding of fundamental structures or processes related to the origins and evolution of the universe, the solar system, the planet Earth, or life on Earth?

A.8 In what ways will the initiative stimulate integration or combination of now separate concepts or information? Will it advance the modeling and theoretical description of natural processes?

A.9 Will the program or initiative yield other achievements?

TABLE A-2. Summary of Potential for New Discoveries and Understanding (Questions A.5 to A.9)

(a) Potential Accomplishment	(b) Potential Significance	(c) Likelihood of Realizing Potential
Advances in knowledge, new phenomena	S_1	L_1
Improved observations or measurements	S_2	L_2
Answers to fundamental questions about structure, evolution, or origins	S_3	L_3
Integration of information; advances in modeling	S_4	L_4
Other achievements	S_5	L_5

Directions: Use the following guidelines to assign points.

Column (b):

Values	Definition
9-10	Exceptional and notable significance
7-8	High potential significance
4-6	Average potential significance
2-3	Some potential significance
0-1	Little or no potential significance

Column (c):

Values	Definition
90-100	Very high likelihood; 90-100% chance of success
70-89	High likelihood; 70-89% chance of success
50-69	Moderate likelihood; 50-69% chance of success
20-49	Some likelihood; 20-49% chance of success
0-19	Not likely; 0-19% chance of success

3. Uniqueness and Special Motivations

> Please provide concise answers to the following questions.

A.10 Are their special reasons for proposing this initiative now? Is a special time schedule or are special facilities necessary for implementing this program or initiative?

A.11 Could the desired knowledge be obtained in other ways? Why is the proposed initiative the most desirable way to proceed?

4. Summary of Scientific Importance of the Proposal

Please provide an overall assessment of the scientific importance of the proposed program or initiative:

        To the proposing discipline(s) _____

        To science generally _____

Directions: On a scale of 1 to 10, evaluate the overall importance of the initiative to understanding in the proposing discipline(s) and to science generally. Note that a score of less than 10 does not necessarily imply that the initiative is not worth undertaking as it may be justifiable on the basis of other criteria assessed subsequently in this evaluation.

Values	Definition
9-10	Critically important; likely to improve or augment understanding fundamentally
7-8	Highly important; likely to improve or augment understanding significantly
4-6	Moderately important; likely to provide incremental understanding or point the way to important advances
2-3	Somewhat important; likely to provide incremental understanding
0-1	Not very important; likely to improve understanding only marginally

SECTION B. CONTRIBUTION TO NATIONAL GOALS

This section of the questionnaire addresses the initiative's contributions to national goals and public objectives, and the likelihood that the initiative will achieve them.

> Please provide concise answers to those questions that are relevant to the initiative; indicate those that are not applicable with "N.A."

B.1 In what ways is this initiative related to broad national goals such as:

- *human welfare (including quality of life, health, safety, etc.),*
- *economic vitality, and/or*
- *national security?*

B.2 Will the results assist society in planning for the future?

B.3 What is the potential for stimulating technological developments that have application beyond this particular initiative?

B.4 How will the initiative contribute to public understanding of the physical world and appreciation of the goals and achievements of science?

B.5 In what <u>unique</u> ways will the initiative contribute to education by generating student interest in science or by attracting students to careers in science or engineering? Distinguish between contributions expected to students in elementary, secondary, undergraduate, and graduate school.

B.6 In what ways will the initiative contribute to international collaboration and understanding?

B.7 Will the initiative contribute to national pride and to the image of the United States as a scientific and technological leader because of the magnitude of the challenge, the excitement of the endeavor, or the nature of the results?

B.8 Will this initiative contribute to the development of commercial space activity?

B.9 What impacts will the initiative have on applications or public services? Will it be possible to improve or curtail any current activities as a result of this initiative?

B.10 Will this initiative have any adverse consequences for science, the environment, or society?

TABLE B-1. Evaluation of Contribution to National Goals (Questions B.1 to B.8)

(a) Potential Benefit	(b) Impact of Initiative (0 to 10 points)	(c) Likelihood of Impact (0 to 100%)
Human welfare		
Economic strength or growth		
National security		
Assistance in planning for future		
Potential for stimulating technological development		
Improvement in public scientific understanding		
Unique contributions to education		
Fostering international collaboration and understanding		
Contribution to national pride and image		
Contribution to commercial space activity		

Directions:

Column (b): Assign points according to the scale below.

Points	Definition
9-10	Very significant demonstrable impact
7-9	Significant demonstrable impact
4-6	Moderate demonstrable impact
2-3	Some demonstrable impact
0-1	Indirect or little impact

Column (c): Indicate the likelihood that this initiative will realize or produce the potential benefit. Your assessment should reflect your best judgment about the resilience of the operation and execution of the initiative and any follow-up actions or technology transfer necessary to produce potential benefit. Use the following scale.

Values	Definition
90-100	Very high likelihood; 90-100% chance of success
70-89	High likelihood; 70-89% chance of success
50-69	Moderate likelihood; 50-69% change of success
20-49	Some likelihood; 20-49% chance of success
0-19	Not likely; 0-20% chance of success

SECTION C. PROGRAMMATIC ASPECTS

This section addresses various aspects of the feasibility, readiness, cost, and risks of the initiative.

1. Feasibility and Readiness: Status and Plans

> Please provide concise answers to the following questions.

C.1 What is the current status of this initiative? What planning has been completed? Who has been involved? What studies have been done? How mature is the concept? How mature is the design of the program or mission? What approvals or endorsements have been granted?

C.2 What are the plans for developing and implementing this initiative? Include descriptions of:
- *Programmatic actions necessary to advance the initiative;*
- *Scientific studies planned or necessary to advance the initiative;*
- *Ways the scientific community will participate in the design and implementation of the program or initiative;*
- *Long-term requirements for special facilities or associated scientific investigations, including launch or on-orbit facilities;*
- *Technological developments required for success;*
- *Plans for an end-to-end demonstration of success;*
- *Plans for processing and analyzing data and for supporting associated research and analysis; and*
- *Any new federally sponsored research activities or institutions required.*

C.3 What is the schedule envisioned for implementing this initiative? What events are planned? What studies will be performed? What are the major milestones for completing the initiative?

C.4 Is there a community of outstanding scientists committed to the success of the initiative? If yes, please indicate how you determined that this is so and how the community interest has been expressed (e.g., NRC report, AAAS session, workshop report, ICSU plan, etc.). In what ways will the community participate in the events and studies discussed in the answer to Question C.3?

2. International Involvement

C.5 Are there scientific or programmatic advantages to cooperating with other nations in pursuing this initiative?

C.6 What fraction of the programmatic support (such as facilities, launch capabilities, or data systems) might advantageously be provided by other nations or international organizations? How critical would that support be to success?

3. Costs of the Initiative

C.7 What will be the total direct cost of the initiative for operating, development, and life-cycle costs in current-year dollars? Use the following ranges:
  a. < $200 million
  b. $200 million to $500 million
  c. $500 million to $1 billion
  d. $1 billion to $5 billion
  e. > $5 billion

How did you derive this cost estimate? What is included and what is omitted? What are the major sources of uncertainty regarding these cost estimates? On what previous experience are these estimates based?

4. Risks and Risk Mitigation

C.8 What are the critical contingencies and risks associated with this initiative? What may impede success? Consider, as applicable, the following:
  - *Technological developments,*
  - *Commitment of an adequate community of outstanding scientists,*
  - *Dependence on international contributions,*
  - *Requirements for data and information systems, and*
  - *Requirements for innovative administrative or program management arrangements.*

C.9 How will the critical risks be mitigated?

Suggested format for response (C.8 and C.9):

Introductory statement about the risks and difficulties of implementing the initiative. The key risks and contingencies for this initiative are:

First critical risk.
  • Description and explanation of the nature of the risk and its potential impact
  • Plan for mitigating this risk

Second critical risk.
  • Description and explanation of the nature of the risk and its potential impact
  • Plan for mitigating this risk

Third critical risk. . . .

C.10 Suppose the initiative is funded but then fails technologically or programmatically (e.g., launch failure, spacecraft power failure, critical instrument failure, withdrawal of international partner). What would be the consequences for the proposing discipline, science in general, the nation?

SECTION D. SUMMARY

D.1 *What relative weights do you believe should be assigned to this initiative's scientific benefits and contribution to national goals (on a scale of 100)?*

Scientific benefits W_1 _____

Contribution to $100 - W_1$ _____
national goals

TABLE D-1. Overall Rating of Initiative

(a) Category	(b) Rating of Benefit	(c) Rating of Benefit Relative to Cost	(d) Rating of Likelihood for Achieving This Benefit
Scientific benefits			
Contribution to national goals			

Directions: Provide your estimates of ratings relative to a hypothetical population of all proposals within the discipline and all proposals within science. Use the following scale:

Rating	Column (b)	Column (c)	Column (d)
9-10	An outstanding and unusual opportunity	Few proposals offer this much value	Few proposals have this much likelihood of success
7-8	An important opportunity in the top 20 percent of all initiatives	Only the top 20 percent offer this much value	Only the top 20 percent have this much likelihood of success
4-6	An average opportunity	Many proposals offer this much value	Many proposals have this much likelihood of success
2-3	Below average opportunity	Most proposals offer more value	Most proposals offer greater likelihood of success
0-1	Little benefit in this category	This proposal offers little value in this category	This proposal has little likelihood of success in this category

Appendix B

Qualitative Evaluation Form

QUALITATIVE EVALUATION FORM

Evaluator:_____

Rank	Name of Initiative	This Set	Relative Rank, Space Science	Relative Rank, All of Science
1	_____	_____	_____	_____
2	_____	_____	_____	_____
3	_____	_____	_____	_____
4	_____	_____	_____	_____
5	_____	_____	_____	_____

(Top)

This Set
1   2   3   4   5

Space Science and Applications
1   2   3   4   5

All of Science
1   2   3   4   5

Appendix C

Quantitative Evaluation Form

EVALUATION FORM

SECTION A. Evaluation of the Overall Scientific Merit of the Initiative

(A1) Are there any responses in Section A that you find particularly compelling? If so, please identify the question and describe why.

(A2) Are there any responses in Section A with which you disagree in whole or in part? If so, please identify the question and describe why.

(A3) Score: Using the scale below, evaluate the overall importance of the initiative to understanding in the proposing discipline(s), to space research, and to science generally. Your score need not agree with the score given by the proposer(s). Note that a score of less than 10 does not necessarily imply that the initiative is not worth undertaking as it may be justifiable on the basis of other criteria assessed subsequently in this evaluation.

Significance to:

The proposing discipline(s) _____

Science generally _____

Significance	Definition
9-10	Critically important; likely to improve or augment understanding fundamentally
7-8	Highly important; likely to improve or augment understanding significantly
5-6	Moderately important; likely to provide some increased understanding or point the way to important advances
2-4	Somewhat important; likely to provide incremental understanding
< 2	Not very important; likely to improve understanding only marginally

SECTION B. Evaluation of the Initiative's Contribution to National Goals

(B1) Are there any responses in Section B that you find particularly compelling? If so please identify the question and describe why.

(B2) Are there any responses in Section B with which you disagree in whole or in part? If so, please identify the question and describe your concern.

(B3) Score: Using the scale listed below, evaluate the overall significance of the initiative in contributing to national goals identified by the proposal and/or modified by your response to question B1.

                Significance   _____

Significance	Definition
9-10	Very significant direct, demonstrable impact
7-8	Significant direct, demonstrable impact
5-6	Moderate direct, demonstrable impact
3-4	Some direct, demonstrable impact
< 20	Indirect or little impact

SECTION C. Evaluation of Programmatic Aspects and Costs of the Initiative

- (C1) Are there any responses in Section C that you find particularly compelling? If so, please identify the question and describe why.

- (C2) Are there any responses in Section C with which you disagree in whole or in part? If so, please identify the question and describe your concern.

- (C3) Score—Programmatic Aspects and Costs: Using the scales below, evaluate the quality of the following programmatic aspects of the initiative:

    Feasibility of initiative   _____

    Reasonableness of cost estimates _____

Feasibility

Score	Definition
9-10	Initiative ready to go as outlined
7-8	Initiative almost ready to go
5-6	Initiative still needs some structure
2-4	Initiative needs significant structure
< 2	Initiative poorly structured

Reasonableness of Cost Estimates

Score	Definition
9-10	Estimated costs appear reasonable
7-8	Estimated costs may be within a factor of 1.5
5-6	Estimated costs may be underestimated within a factor of 2
2-4	Estimated costs may be underestimated, but by less than an order of magnitude
< 2	Estimated costs may be underestimated by greater than an order of magnitude

SECTION D. Evaluation of Benefits and Likelihood of Success

Assessment of Benefits

(a) Category	(b) Proposed Weighting	(c) Rating of Benefit Relative to Other Scientific Initiatives
Scientific merit and objectives (Section A)	W_1	
Contribution to national goals (Section B)	$10-W_1$	

Directions:

Column (b): Specify the relative weighting of each of the two categories of benefits you believe should be used in evaluating this initiative.

Column (c): Use the scale below to rate each item (independently of weights).

Rating	Definition
9-10	An outstanding and unusual opportunity; few proposals offer this much potential or this much benefit
7-8	An important opportunity; in the top 20 percent of all initiatives
5-6	An average opportunity
2-4	A below average opportunity
< 2	Initiative offers little benefit in this category

Assessment of Benefit-to-Cost Ratio

(a) Category	(b) Proposed Weighting	(c) Rating of Benefit-to-Cost Ratio Relative to Other Scientific Initiatives
Scientific merit and objectives (Section A)	W_1	
Contribution to national goals (Section B)	$10-W_1$	

Directions:

Column (b): Specify the relative weighting of each of the two categories of benefits you believe should be used in evaluating this initiative.

Column (c): Use the scale below to specify ratings for each item (independently of weights)

Rating	Definition
9-10	An outstanding and unusual opportunity; few proposals offer this much potential or this much benefit
7-8	An important opportunity; in the top 20 percent of all initiatives
5-6	An average opportunity
2-4	A below average opportunity
< 2	Initiative offers little benefit in this category

(D1) Using the scale below, rate the likelihood of the initiative's achieving the benefits within the time frame and costs proposed.

Likelihood of success:

Scientific merit and objectives _____

Contribution to national goals _____

Likelihood	Definition
9-10	Few proposals have this much likelihood of success
7-8	Only the top 20% have this much likelihood of success
5-6	Many proposals have this much likelihood of success
2-4	Most proposals have this much likelihood of success
< 2	This proposal has little likelihood of success

In the table below, give your assessment of the likelihood that the initiative will be successful in meeting its scientific objectives and in providing the social benefits envisioned by the proposers.

Assessment of Likelihood of Success

(a) Category	(b) Likelihood of Success
Scientific objectives	
Contribution to national goals	

Directions:

Column (b): Use the scale below to specify likelihood of success.

Likelihood	Definition
90-100	Very high likelihood; 90-100% chance of success
70-89	High likelihood; 70-89% chance of success
50-69	Moderate likelihood; 50-69% chance of success
20-49	Some likelihood; 20-49% chance of success
< 20	Not likely; 0-20% chance of success

Appendix D

Notes on Quantitative Evaluation Schemes

The objective of a quantitative evaluation scheme in a priority-setting process is to assign a numerical measure of overall relative merit to each initiative within a category of candidates. Such schemes have the advantage of averaging over many individual judgments, of permitting sensitivity tests to be performed, and of providing complete traceability. They have the disadvantage that the uncertainty associated with costs and benefits and likelihood of success makes them inaccurate if such values are used directly or appear in ratios, and thus the apparent precision of numerical evaluation schemes can be misleading and open to misinterpretation.

The purpose of this appendix is to outline the main features of quantitative evaluation schemes as an aid to those who might desire to use them and to suggest an approach that, to some degree, appears to mitigate the effects of uncertainty about numerical quantities such as costs.

THE BASIC CONCEPT

To illustrate the basic concept, let $i = 1, 2, ..., N$ denote the candidates, let S_{ij} for $j = 1, 2, ..., J$ denote the numerical score associated with the ith candidate and the jth criterion, and let W_j for $j = 1, 2, ..., J$ with

$$\sum_j W_j = 1 \tag{1}$$

denote the relative weight assigned to the jth criterion. The priority rank P_i for each candidate can be obtained from

$$P_i = \sum_j W_j f(S_{ij}) \qquad i = 1, 2, ..., N \tag{2}$$

where f is a scaling function, usually taken to be the identity for a linear combination of criterion scores and weights. In this case

$$P_i = \sum_j W_j S_{ij} \qquad i = 1, 2, ..., N \tag{3}$$

To rank the initiatives, each evaluator assigns a criterion score S_{ij} for each criterion for each candidate. Averaging the elements of the matrix across evaluators thus produces a matrix of average criterion scores to use in (2) and (3). The set $\{P_i\}$ then determines the ranking or priority order of the candidates.

DETERMINING THE WEIGHTS

The weights in (2) and (3) are determined by the evaluation group or specified in advance by the organization receiving the recommendations. They should be independent of the criterion scores. One way to proceed[1] is for the group to use one of the criteria as a basis and scale the others to it. Thus the most important criterion might be selected as the basis and the relative importance of each other criterion assigned individually in comparison to the basis. Thus if $W_2 = aW_1$, $W_3 = bW_1$, and $W_4 = cW_1$, then (1) implies that $W_1 = 1/(1 + a + b + c)$ and hence all the weights are determined.

As a specific example based on the criteria discussed in Chapter 2 of the main report, the group might decide that scientific merit and cost are equally important so that $W_M = W_C$, that scientific merit is four times as important as contributions to national goals so that $W_M = 4W_G$, and that cost is twice as important as likelihood of success so that $W_C = 2W_L$. Then the weights will be $W_M = W_C = 4/11$, $W_L = 2/11$, and $W_G = 1/11$.

DEFINING THE CRITERION SCORES

The validity of a quantitative evaluation scheme depends on defining criterion scores that accurately and reliably measure the relative rating of the candidate initiatives with respect to the criteria. Moreover, the scores must be suitably scaled if scores and weights are to be intercompared. To see this, consider an extreme case where scientific merit is measured on a scale of 1 to 10 and cost in dollars: for space research flight missions, the cost will dominate the estimates (2) and be given a weight much greater than intended.[2]

The scaling issue is critical in many contexts. For example, in benefit-cost analysis, there is an extensive literature on ways to estimate the dollar value of social benefits of various kinds so that the benefit-cost ratio is dimensionless when costs are estimated in dollars.

Because it can be difficult to assign an independent dollar value to increases in scientific understanding or even to the social benefits of scientific progress, a different approach is suggested that uses an indicator function to translate both subjective and numerical estimates into integers 1 through 5. The relation between the integer scores and the definitions used in the template and evaluation forms (Appendixes A and C) are shown in Table I.

The translation of cost ranges within a category to an integer cost criterion score completes the scheme. Let the lower limit of cost in a category be L_1 and the upper limit be L_2. The assignment of scores can now logically be based on the expected distribution of costs within the category. Let x be

TABLE I Relation of Criterion Scores to Definitions from Templates and Forms in Appendixes A and C

	Criteria		
Score	Scientific Merit	Social Benefits	Likelihood of Success
5	An outstanding and unusual opportunity; few proposals offer this much potential	An outstanding and unusual opportunity; few proposals offer this much benefit	Very high likelihood; 90-100% chance of success
4	An important opportunity; in the top 20% of all initiatives	An important opportunity; in the top 20 percent of all initiatives	High likelihood; 70-80% chance of success
3	An average opportunity	An average opportunity	Moderate likelihood; 50-70% chance of success
2	A below average opportunity	A below average opportunity	Some likelihood; 20-49% chance of success
1	Initiative offers little benefit	Initiative offers little benefit	Not likely; 0-20% chance of success

NOTE: All statements are to be interpreted relative to the specific category of candidates under consideration.

a numerical variable such as cost or probability of success and let x have a probability distribution $F(x)$ within the category under consideration. Then integer scores S may be assigned according to the scheme $S = 1$ for $0 \leq F(x) \leq 0.2$, $S = 2$ for $0.2 < F(x) \leq 0.4$, ..., $S = 5$ for $0.8 < F(x) \leq 1.0$. The corresponding values of x are shown for both uniform and Gaussian distributions in Table II, along with the class limits for cost in a category with L_1 = \$100M and L_2 = \$900M. These two distributions were chosen for illustrative purposes; others are feasible and may be preferable.

A BENEFIT-COST ESTIMATE

The four major criteria specified in Chapter 2 suggest the construction of a single measure to assess the relative value of initiatives. Using the integer criterion scores already defined, let M be the score for scientific

TABLE II Class Limits Corresponding to Integer Criterion Scores for Numerical Variables x

Score	$F(x)$	Values of x		Class Limits for Range from $100M to $900M (in $ M)	
		Uniform	Gaussian	Uniform	Gaussian
	1.0	L_2	L_2	900	900
5					
	0.8	$0.2L_1 + 0.8L_2$	$0.36L_1 + 0.64L_2$	740	610
4					
	0.6	$0.4L_1 + 0.6L_2$	$0.46L_1 + 0.54L_2$	580	530
3					
	0.4	$0.6L_1 + 0.4L_2$	$0.54L_1 + 0.46L_2$	420	470
2					
	0.2	$0.8L_1 + 0.2L_2$	$0.64L_1 + 0.36L_2$	260	390
1					
	0	L_1	L_1	100	100

NOTE: For the Gaussian distribution, L_1 is assumed to correspond to a standard variable of -3, L_2 to 3.

merit, G for contributions to national goals, C for cost, and L for likelihood of success. For weighting, let scientific merit be a times as important as contributions to national goals so that $W_M = aW_G$. Then, on the argument that since benefits are not certain, they should be discounted relative to the probability of obtaining them, we can define a single measure P of merit within a category as

$$P = \frac{(MW_M + GW_G)L}{(W_M + W_G)C} = \frac{(aM + G)L}{(1+a)C} \quad (4)$$

While the benefit/cost ratio is typically used, a preferred measure is net benefit, or the difference between benefits and costs. To see this, we can define an alternative measure P_N for the case already considered as

$$P_N = \left[\frac{(aM + G)L}{1+a}\right] - C. \tag{5}$$

The advantage of the net benefit measure is that it is invariant with respect to definitional issues associated with benefits and costs. For example, costs can be considered negative benefits or "disbenefits." Costs defined as disbenefits change the value of the benefit/cost ratio by reducing the numerator, compared with costs that are categorized as costs, which increase the denominator. Whether costs are disbenefits or costs per se does not alter the net benefit measure, however.[3]

These measures might also be used to compare initiatives in different cost categories provided that the assessments of benefits were made relative to the hypothetical population of *all* initiatives regardless of cost category and provided that the cost indicator is adjusted to cover the range spanned by all the categories. Whether a collection of less costly initiatives would then be favored relative to a single expensive initiative would depend on whether the evaluations of scientific merit were correlated with cost.

NOTES

1. Institute of Medicine, *Setting Priorities for Health Technology Assessment—A Model Process*, Committee on Priorities for Assessment and Reassessment of Health Care Technologies, National Academy Press, Washington, D.C., 1992.

2. The procedure proposed by the reference in footnote 1 suffers from this disadvantage since very different variables are combined without scaling.

3. A typical example is travel congestion associated with expanding a highway. The congestion can be considered as a disbenefit, and its value subtracted from benefits, or a cost, and its value added to costs. The benefit/cost ratio, but not the net benefit measure, will be sensitive to the definition of congestion.

www.ingramcontent.com/pod-product-compliance
Lightning Source LLC
Chambersburg PA
CBHW081733170526
45167CB00009B/3798